I0482193

I have been aware of the concepts that inspire this book for many years. It has been frustrating for me to fail again and again at finding some method that could prove or advance this idea. We currently do not have the science or technology to confirm these concepts. What did the first person do who realized that the stars were not pinholes in some giant bowl. That the sun was not a personality. That human sacrifice did not bring rain.

I expect these explorers of ideas and theories talked about, wrote about, debated about the ideas. So let's get started.

PLANET EARTH AND THE "GAIA" CONCEPT

John Lovelock is a British scientist who proposed that Planet Earth is a huge organism.

 He documented that various qualities of the biosphere are constantly being regulated to sustain and maximize life forms even though these qualities would not occur by chance. He demonstrated that vast groups of microscopic organisms work together somehow to create this perfect environment for all manner of life forms to exist and flourish. Global temperature, the composition of the atmosphere, and the levels of salt in the oceans are three major examples of this.

Lovelock realized that the entire planet seemed to be functioning in a coordinated manner. Like a huge organism.

All things living are the product of small things reorganizing themselves into larger things of increasing complexity. Nature demonstrates this to us in the infinite variety of life and life forms on our planet.

Human beings are a compilation and synthesis of many different groups of cells who have organized themselves into organ systems.

These organ systems at a higher level produce a unified autonomous organism. A human being.

So Lovelock has documented that our planet is functioning in a globally coordinated fashion. Planet earth is composed of innumerable small systems of life forms.

 And we know that smaller systems tend to reorganize themselves into larger systems that are autonomous.

Dr. Lovelock and his colleague Lynn Margolis stopped short of following these concepts to the logical conclusion, in my opinion. They likened Planet Earth to a giant one celled organism.

I propose in the following pages that this really isn't the appropriate perspective. Why would we think that this giant organism would resemble a more simple organism at its highest state?

Remember, smaller life form systems group themselves in order to form larger systems that are more complex and autonomous.

PLANET EARTH AS AN ENTITY

So if Planet Earth is an entity we would be expecting to see qualities involving respiration, circulation, and brain functioning.

We have lava boiling up from the Earth's core in the branching patterns we see everywhere in nature. Are these the conceptual blood vessels?

Or would it be water for circulation?

Rivers and streams as the vessels and capillaries?

The circulation pattern of clouds and rain-evaporation and condensation-as respiration?

Hard to be definitive but pretty clear that circulation and respiration are happening. That is clear, don't you think? Lots of ebbing and flowing going on , in circular and cyclical patterns .

But Earth with a brain? That's hard to see.

Our brain is an organ characterized by electro-chemical activity.

A complex compilation of electrical impulses and brain chemistry.

Several different neuro-transmitting chemicals.

All these characteristics exist also in the Earth's atmosphere.

 We think air is just a bit of wispy nothing but it is actually on the continuum of fluid.

Meteors hit the atmosphere with such force that they burn up.

The atmosphere is a membrane for the most part.

Lightning strikes the Earth 55,000 times a day.

Not commonly known that lightning radiates up in to the atmosphere also in what are called sprites and blue angels.

 This electrical energy radiates out in to the different levels of the atmosphere. Radiates more up than down really.

Remember your science? All the atmospheric levels-troposphere, stratosphere, ionosphere, etc.

And there are several types of cosmic particles streaming and bombarding the atmosphere. All manner of energy and interaction happening from the molecular to the subatomic.

Most not visible to the naked eye. Making a brain.

Earth's Brain

A few years ago I had an experience where I was sitting in the sun on a nice day and I noticed a colony of ants near me. I watched them for a bit. And I had a revelation.

To me, the ants seemed to be frantically scurrying to and fro.

There was no alarm in their situation. They had no awareness that I was nearby watching. But if you watch ants they cover ground 10 or 20 times their body length in just a second.

If a human would cover that many body lengths in that brief instant I think we would be traveling 60 miles an hour or something shocking. So that was interesting.

But I also realized that the ants were not hurrying. Why would they be? What animal or life form is rushing, running, sprinting, when taking care of normal day to day activities?

Those ants were walking.

And they only seemed to be rushing to me because I am so much bigger.

Think about an elephant and a mouse. If you have ever chased a mouse you know they are lightning quick. Think how quick they would seem to an elephant.

Think about a hummingbird. You could never quickly touch a hummingbird. I think our movements would be in extreme slow motion for that high speed little bird.

The very important point that is connected to these examples is that larger animals experience time more slowly than smaller animals.

Smaller organisms move and react more quickly because they are living and experiencing more quickly. And their nerve impulses have much less distance to travel.

One second to me is maybe 3 seconds to a mouse? Or what would it be subjectively?

So the EARTH is HUGE. And we are infinitesimally tiny compared to this huge organism. So one second to the Earth would be how long? In a human beings terms?

As close as I can figure, one second to the Earth is 7 hours to us. I got that with complex ratios between the size of ants to mice to humans to elephants-using Mass of the life forms as the unit of measurement and comparison.

I could be way off-impossible to test or confirm. But it's a lot more than an elephant to an ant. The disparity in actual size and Mass. You are suddenly working with numbers with 70 zeros behind them.

This is why we don't perceive Earth as a life form-it's too Huge. And it is subjectively experiencing time MUCH more slowly than us. A twenty four hour period for us may only be one second of consciousness for Planet Earth. Mites that live in our eyelashes would not perceive us as a unified entity. We would be it's vast world. No possible way of perceiving each other. We don't know the mites exist without a microscope.

So there you have it. Planet Earth as a thinking conscious entity. What sort of consciousness would this entity possess? Impossible to say but it would be almost infinitely more evolved than a human being.